D1348447

The
PRICE GUIDE
to
VICTORIAN SILVER

by

Ian Harris

of N. Bloom and Son Ltd.

Published by   THE
ANTIQUE
COLLECTORS'
CLUB

Printed in England by
Baron Publishing, Church Street,
Woodbridge, Suffolk.

FOREWORD

The Antique Collectors' Club

The Antique Collectors' Club, formed in 1966, pioneered the provision of information on prices for collectors. The Club's monthly magazine *Antique Collecting* was the first to tackle the complex problems of describing to collectors the various features which can influence prices. In response to the enormous demand for this type of information the *Price Guide Series* was introduced in 1968 with **The Price Guide to Antique Furniture**, a book which broke new ground by illustrating the more common types of antique furniture, the sort that collectors could buy in shops and at auctions, rather than the rare museum pieces which had previously been used (and still to a large extent are used) to make up the limited amount of illustrations in books published by commercial publishers. Many other price guides have followed, all copiously illustrated, and greatly appreciated by collectors for the valuable information they contain, quite apart from prices.

Club membership, which is open to all collectors, costs £6.95 per annum. Members receive free of charge *Antique Collecting*, the Club's monthly magazine, which contains well-illustrated articles dealing with the practical aspects of collecting not normally dealt with by magazines. Prices, features of value, investment potential, fakes and forgeries are all given prominence in the magazine.

In addition members buy and sell among themselves; the Club charges a nominal fee for introductions but takes no commission. Since the Club started many thousands of antiques have been offered for sale privately. No other publication contains anything to match the long list of items for sale privately which appears monthly.

The presentation of useful information and the facility to buy and sell privately would alone have assured the success of the Club, but perhaps the feature most valued by members is the ability to make contact with other collectors living nearby. Not only do members learn about the other branches of collecting but they make interesting friendships. The Club organises weekend seminars and other meetings.

As its motto implies, the club is an amateur organisation designed to help collectors to get the most out of their hobby; it is informal and friendly and gives enormous enjoyment to all concerned.

For Collectors   —   By Collectors   —   About Collecting

The Antique Collectors' 5, Church Street, Woodbridge, Suffolk

CONTENTS

INTRODUCTION

After the *Price Guide to Antique Silver* was published in September 1969, it became clear that in so comprehensive a survey of silver available to collectors, it had been impossible to treat the Victorian period in the detail it deserved.

Taking in the short reign of William IV, the period lasts from one dividing line, the end of the Georgian period, to another, the beginning of this century. It was a period which saw enormous social changes, with the consolidation of a prosperous middle class, and an increase in the amount and distribution of wealth amongst it.

Until recently, all art and artefacts of this enormously productive period were scorned by serious collectors; now, however, we look at them with a new eye. For many pieces of this period form a modest introduction to the fascination of collecting.

This, therefore, is why we thought Victorian silver deserved a price guide to itself. There are so many more things to be found, so many odd objects. The price levels, too, are so much lower, and are likely to appeal perhaps to the less experienced collector who, we hope, will find this book especially useful.

This is intended as a *practical* guide to quality, condition, and value. I have made little effort to go into the immense amount of information available on designers and manufacturers, schools of thought or influence.

All these are certainly of great interest, and can be found in the increasing number of books written by those with the time and inclination for research. I hope you will read this practical guide in conjunction with them, and learn something from all of us.

October 1971.

PREFACE

Basis of Price

As in the Guide, the prices given are those prevailing between dealer and dealer, and should be in the region of saleroom prices. This I think more realistic than the retail prices sometimes given. The private buyer may expect to pay somewhat more than the higher price when buying, and receive around the lower price if selling. An exceptionally wide price margin usually means that rather similar looking articles vary widely in size and/or quality.

A Word About Auctions

Many people believe that by buying in auction rooms they can cut out the dealer and his profit. To a point, this is true. Most dealers will not go substantially above their own valuation in order to keep out private buyers. There are several dangers, however. Firstly, you may well be bidding against another private buyer who doesn't understand the value either. Dealers speak in awe of some really stupendous prices when this has happened. Secondly, you may be bidding for a poor, damaged, faked, repaired, or otherwise undesirable article, which no decent dealer would stock, or permit you to buy. This is even worse than over-paying, since whatever you pay for a bad article is too much. Now, however, I am going to try to tell you what to look for, so you won't even stand the risk!

What to look for

Even if you habitually buy only from dealers of established reputation and authority, it is nice to know that in your own mind you agree with what they tell you. But many collectors do not have easy access to such dealers, and many enjoy buying at auction sales or from their local dealer, who, because he is unable to specialise, is unlikely to have a really sound knowledge.

The main faults to be found with silver are wear, eventually necessitating repair; legitimate alterations, which do not pretend to be other than they are; and deliberate faking, which also usually involves alteration with a view to improving the desirability of an article.

The most usual fault is wear. Silver was made to be used, and it is only since the beginning of this century that it was considered to have much antiquarian interest. By looking at lots of silver, you will soon be able to see whether the decoration is crisp and sharp or not. Engraving will look a little faded, and the edges of the engraved lines will no longer be sharp. With embossing, the high spots will have lost all detail, and present a smooth and shining appearance. Much Victorian silver is embossed in quite high relief. Where this is done, it stretches the silver, making it thinner, and the more pronounced the decoration, the thinner it is. At the same time, these high spots receive the most wear from cleaning. Therefore, all embossed decoration should be examined most carefully for small holes. These of course can be filled in with either hard or soft solder, but usually the latter. This is not much different from the stuff the plumber uses, unlike hard solder, which is almost as pure silver as the article itself. Hard solder, like silver, only melts at a very high temperature, and is consequently much more difficult to use. Lead solder tends to corrode gold and silver, as well as being very soft and lacking in strength, and having a very low melting point. The characteristic dark grey colour of soft solder is very evident, especially looking from the inside of a piece, unless it is hidden, say, by the coat of tannin inside a teapot, or by plating over.

This last is more difficult to spot, but, like most things which have interfered with the natural patina, it tends to leave the article looking rather pale and white, with a surface gloss very different from the hard, deep shine which develops on an un-restored piece. Sometimes a plated article also displays signs of porosity, and of course soldered seams (where you should be able to see the faint line of silver solder) would also be plated over and disappear.

You may have seen dealers breathing on an article. This dulls the surface shine, so that joins and repairs, as well as hall-marks, are shown up much better.

The high spots on any piece are the first places to look for wear. Some types of fluting, where it terminates in a "V" edge, are particularly prone to damage, and are difficult to repair. The handles of teapots and tankards are often pulled from the body; always inspect the junctions of handles and feet from the outside, and more especially the inside. Interestingly, the teapot on a four-piece set is often very much more worn than the other pieces, as it was used several times daily, whilst the others had little use at all.

Another main hazard is the unskilled removal of engraving. Even now, most dealers will automatically remove a presentation inscription or initials from a piece of silver before putting it on sale. If unskilfully done, this may make the piece very thin. At best, it will spoil the patina. Usually, when engraving is removed, it will leave a hollow either inside or outside, front or back, especially in the case of salvers. The engraved border surrounding the inscription seldom escapes altogether, especially the surrounds to the small panels on snuff boxes and vinaigrettes. I do not say that removing these inscriptions necessarily *reduces* the value; in most cases it will make the article more saleable. What is important is that you should realise when it has been done, and look further to see whether serious damage has resulted.

The Victorians were fond of altering earlier pieces. 18th century baluster tankards were a particular favourite. They took these lovely plain pieces of silver, and added a lip. They then cut the handle in two places, and inserted ivory insulators. Finally, they embossed the whole thing with farmyard scenes or flowers. They embossed all plain articles; nothing was too good to be ruined. This habit, however, mainly concerns the collector of Georgian silver, although such altered pieces must be considered as Victorian as far as price is concerned.

Deliberate faking is fairly uncommon since it is only in the last few years that Victorian silver could conceivably have been worth faking. Around the turn of the century, a number of poor fakes were made by transposing the part of the stem of a spoon or fork with the hall-mark into one side of a square base; but there again, such fakes were copies of Georgian, not Victorian, pieces. Actual forgery of hall-mark punches is so rare as to be hardly worth considering.

However, unlike Georgian silver, Victorian, even now, is not really worth faking to any great extent, and even less so during the last fifty years, so it's the wear and repairs that are the first concern.

In the *Antique Silver Guide,* I went at length into the question of engraving, particularly armorial engraving. This has little effect on the value of Victorian silver, but I should still say that an inscription of monogram, unless of historical interest, would *reduce* the value of a piece by 10–15%. As collectors become more enlightened, this may change.

I need hardly say that hall-marks should be clearly legible, and each separate piece should bear a mark. A small thing like the top of a pepper will probably have just the lion passant and maker's mark; larger subsidiary parts like a soup tureen liner or cover will have full marks less one, usually the leopard's head if London. Fruit bowls are sometimes made from soup tureen liners — if so, they'll have a mark missing.

Don't be taken in if someone tells you that, as it's old, you've got to expect a certain amount of wear. There are many 15th century pieces of silver in existence in pristine condition, and you certainly don't have to accept anything less from the 19th! In fact, whether you're a collector or investor, it pays to stick to only the best. In the former case, you'll derive much more satisfaction from your collection, and won't constantly want to swap it around as your discrimination increases, and in the latter it'll show a much better return when you sell it! It is far, far better to buy a small good piece, than a larger but poor one.

Famous Makers

There are no makers of Victorian silver comparable to those earlier makers whose work fetches many times that of their contemporaries. Paul Storr, it is true, survived into the Victorian period, and his mark may double the value of an article. Charles Fox is a maker for whom there is a small band of collectors, and this has been sufficient to increase the value of his work up to 50% for pieces of an unusual character. Charles working alone was followed by the partnership of Charles and George Fox, and finally George Fox alone. All their work is of excellent quality. Other excellent makers, or companies, were Storr and Mortimer, who employed Storr, John Hunt and Benjamin Smith, succeeded by Hunt & Roskell; various combinations of Barnards — the firm is still in existence today; and Garrard — usually for very fine copies of 18th century silver. Elkington & Co. were a highly progressive firm who developed electro-plating, and gilding, but also produced some fine silver, employing some distinguished designers. Their work was hall-marked in Birmingham, where they originated, or London. But there were many other fine craftsmen; in fact, most London-made silver of the 19th century is excellent. Technically, the silversmiths were very advanced, and the cost of workmanship was low. But it is really only Storr, and possibly Fox, who would increase the value of an article. Otherwise one would judge far more on quality, condition, and attractiveness.

One other maker should be mentioned — Nathaniel Mills, the Birmingham vinaigrette, box, and card-case maker. Birmingham since the late 18th century was the home of the "small worker", the button and buckle maker, the thimble maker, etc. The work was very largely mechanised, with beautifully executed die-stamping. Even much engraving was largely acid-etched, possibly with some hand-finishing. Nathaniel Mills's work is not superior, as far as I can see, to that of any other firm, but he was rather more prolific. Perhaps he sounds more of an individual silversmith than other Birmingham firms, most of whom were styled "& Co.", or were partnerships. Another Birmingham maker people have heard of is George Unite, who is somewhat later in date than Mills. Personally, I would give little or no more for these two gentlemen — although I recognise there's a possible selling point — but would again judge for myself the quality of the piece.

Tea and Coffee Sets and their Constituents

For most of the 18th century, coffee pots, teapots and the other pieces of a teaset were considered separate items. With few rare exceptions, the concept of a matching service did not develop until about 1785. By the beginning of the Victorian period, however, the set was universal, and even though one may see individual pieces for sale, they almost all at one time formed part of a set. Exceptions may be the odd "pumpkin" teapot, or other extravagance.

Most of the separate pieces of a tea and coffee set sell perfectly well on their own, with the exception of sugar basins. I shall, therefore, give the prices of the pieces separately, as well as that of complete sets. When buying, careful attention must be paid to condition. It is not unusual to find that the teapot in a set is almost worn out, whilst the rest of the set is in pristine condition, since the teapot may have been used daily for a hundred years, whilst the other items were got out once a month.

Vertical fluting finishing in a 'V' edge is particularly liable to damage, and particularly difficult to repair. Signs may be tiny pinholes along the angle caused by soldering. Embossed sets may have holes in them filled with lead solder, and of course bodies may be thin where arms of inscriptions have been erased. A rather white and porous surface is always suspicious, as it may have been caused by plating over to hide some repair.

It is common to find services in which the pieces have been matched up, the whole being considerably more valuable than the sum of the parts. If all pieces are by the same maker and within following years, the price will be affected only about 5%. If the spread of years is much greater, or if there are three or four dates, the value may decline 10%—15%. Some firms produced the same pattern for many years, and dealers today can often make up sets from pieces brought separately. It is very important to look carefully at odd-date sets to see if the pieces are identical — points of difference usually occur in the spouts, feet, and handles; finials can easily be altered.

Sets in which pieces are by different makers really have little more value than that of the individual pieces. However, they are not common, as pieces seldom match up with those of other silversmiths.

Classic melon pattern tea and coffee set, with pumpkin finials and shell and scroll feet. One of the most popular patterns, either plain, or chased with acanthus on the shoulders. From late Georgian to 1840.

4 piece set	*£450 – £600*
3 piece set	*£150 – £200*
teapot	*£ 60 – £ 80*
coffee pot	*£100 – £125*
cream jug or sugar basin	*£ 25 – £ 35*

A variation on the true melon pattern, made during the same period. Where the flutes come to a sharp edge, look carefully for repairs.

4 piece set	*£400 – £550*
3 piece set	*£140 – £200*
teapot	*£ 55 – £ 75*
coffee pot	*£ 90 – £115*
cream jug or sugar basin	*£ 25 – £ 35*

A coffee pot from a very good type of set circa 1835-1845. The outline is fairly plain, with discreet and "un-lumpy" decoration.

4 piece set	*£450 – £600*
3 piece set	*£150 – £200*
teapot	*£ 60 – £ 80*
coffee pot	*£100 – £125*
cream or sugar	*£ 25 – £ 35*

A magnificent coffee pot of circa 1835. The quality and crispness of
the decoration is superb. This is the standard to compare everything
else!

4 piece set	*£550 – £750*
3 piece set	*£200 – £275*
teapot	*£ 75 – £100*
coffee pot	*£150 – £200*
cream or sugar	*£ 35 – £ 50*

A smallish squat pattern embossed teapot circa 1835. If well-worn
examine the embossing carefully for repairs or soft solder.
£40 – £60

An "apple" teapot by the celebrated makers Charles and George Fox. Teapots of this type, normally of fairly small size, were made in the form of melons, pumpkins etc. They date usually from 1835-1850, although I have seen late Georgian examples. They are popular, and fetch a fair price.

£125 – £175

A so-called Louis XV pattern service, another very popular pattern. Points are the bird-headed masked spouts, scroll feet and mounts, and eagle finials. From 1845-1860. This is a type that is often matched up, because the design was so standardized. This set has a matching kettle, which virtually doubles the value. Sometimes the kettle is made to match, but in plate, originally, to save expense. This would add about £100 to the value of a 4 piece set.

4 piece set	£450 – £550
3 piece set	£140 – £200
teapot	£ 55 – £ 75
coffee pot	£ 90 – £125
cream or sugar	£ 25 – £ 35

An early engraved and panelled service of the best type, about 1845. Although embossing was by far the more common form of decoration, some engraved sets were made, but are marginally less popular.

4 piece set	*£400 – £500*
3 piece set	*£125 – £175*
teapot	*£ 50 – £ 75*
coffee pot	*£ 90 – £120*
cream or sugar	*£ 25 – £ 35*

A miniature tea and coffee set made in Dublin in 1851; you can tell it is not real from the odd proportions. A rarity.

£90 – £120

Coffee pot from an engraved set circa 1860. The complicated outline has disappeared, leaving a coffee pot which is much easier to make. Easier or no, the simpler outline was more fashionable. Lots of engraving.

4 piece set	*£375 – £450*
3 piece set	*£100 – £150*
teapot	*£ 45 – £ 65*
coffee pot	*£ 80 – £110*
cream or sugar	*£ 25 – £ 30*

Three pieces about 1865. Note the low outline of the lid, which is typical of this date. The cream jug does not match the other two pieces, but might be taken to at a casual glance — so be careful.

4 piece set	*£350 – £425*
3 piece set	*£110 – £140*
teapot	*£45 – £60*
coffee pot	*£80 – £100*
cream or sugar	*£25 – £30*

An extravaganza. Note, however, how closely the basic shape of the set matches the previous one. Most of these exotic designs come only as three piece sets, and I price it accordingly.

£135 – £175

A complete contrast. This teapot, made circa 1870 in the Dutch style of a hundred years earlier, shows what good copyists the Victorians were.

£45 – £60

One of the most common types of set of the 1870's and early 1880's. It is known as a can (shaped) set, the essentials being the straight tapering form and flat base. It can be of simple oval outline, or shaped, as here, which is preferable. This has never been a popular type of set, and the price would depend largely on the quality and condition, especially of the engraving, which is the most attractive thing about it. This set is on the better side.

4 piece set	£200 – £300
3 piece set	£70 – £110
teapot	£40 – £60
coffee pot	£60 – £80
cream or sugar	£20 – £25

Another typical set of the 1870's and 1880's, definitely more popular than the previous type.

4 piece set	*£250 – £350*
3 piece set	*£100 – £150*
teapot	*£45 – £70*
coffee pot	*£70 – £100*
cream and sugar	*£25 – £30*

An extremely fine "Bachelor" three piece teaset with engraved decoration, partially gilt. Date 1883. More *objets de vertu* than objects of use, they fetch a lot of money.

£150 – £200

A "Bachelor" three piece teaset with engraving in the Japanese taste
fashionable in the 1880's. The teapot holds about two cups, and the
other pieces are in proportion. They are often of exceptionally interest
ing design and fine quality — perhaps because they were intended for
the refined man of taste and means rather than the average bourgeoi
family. Because of their small size and considered impracticability
they are usually most reasonable in price.

3 piece set   *£45 — £65*

This very unusual bachelor teaset and kettle was designed in the 1880's
as a reaction against fussy Victorianism, to which the Oriental influence
was another reaction. The set was probably designed by Dr. Christopher
Dresser, since its clean lines, functionalism, and rather unusual appearance are typical of his work. Although interesting, it is rather a special-
sed taste, and as yet fetches little money.

£50 – £65 complete

Complete degeneration of taste in the 1880's and 1890's. Lightweight spun bodies (you can judge the gauge from the un-mounted edges of the sugar and cream.) The decoration is over-elaborate, although technically well executed. Unfortunately, it will be as valuable as the previous example.

£50 – £65

An elaborate early Victorian tea-kettle with a stand containing a spirit
lamp to keep the water boiling at the tea table. This is a "Gypsy" type
stand where the kettle swings on pins each side; usually the kettle had
hinged joint at the front, as will be seen on other examples. These
highly decorative kettles are very popular.

£250 – £350

Typical embossed kettle circa 1845-1860, showing the usual method of hinging. Again, a very popular type.

£250 – £350

A later example of engraved decoration, circa 1860-1870. Although an excellent example of its type, with crisp, fine engraving, it is much less desirable than the previous one.

£175 – £250

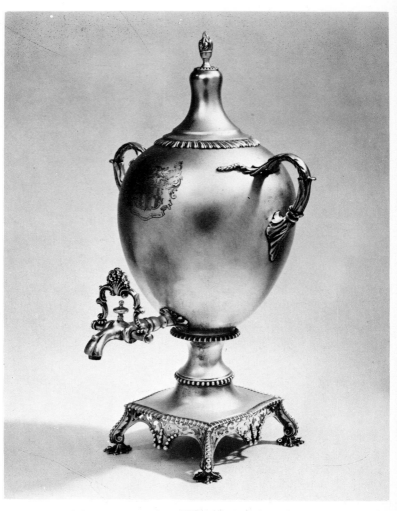

A tea urn. These are unusual in Victorian silver, and this one is a copy of a Georgian one of circa 1770, although the pattern of the tap and the style of arms show it is not.

£300 – £375

Between the 1870's and the end of the century, a number of smallish covered jugs, about 5 or 6 inches high, were made. They were intended for something hot, since they have ivory insulations in the handles; coffee, or hot milk, perhaps. They do not seem to have been part of sets. This is a good example, embossed in the "Chinese" taste (compare with the "Japanese" taste of page 18. Chinoiserie has Chinese figures, phoenixes, pagodas, and "icicles"; Japonaiserie has bamboos and storks, and is usually engraved.)

£50 – £90

An engraved one. I am not sure if it is Chinese or Japanese — it is rather willow-patterny.

£50 – £75

A plain example in the style of a George II shaving jug. It is oval in section. Although made in 1871, it could be earlier or later. Very good quality.

£60 – £95

Most cream jugs in the 19th century were made as part of teasets. This
example is a copy of a Georgian one of circa 1760. Since it is a copy,
the date could be anything between 1850 and 1890 — there is no
style to tell us.

£18 – £25

A typical 19th century novelty cream jug in the form of a Harpy. Cast, and of excellent quality.

£70 – £85

CLARET AND WINE JUGS AND EWERS

Whilst occasional Georgian silver-mounted glass wine jugs are found, they are unusual, and even any other sort of decanter-sized jug is rare, with the exception of so-called beer jugs, which are more like water pitchers than anything else. In the first half of the 19th century, most wine jugs were all metal, although glass ones are found. After about 1860, however, all-metal ones are unusual, but silver-mounted glass ones proliferate. Considering the number that must have got broken, it's surprising how many remain, mostly from 1880-1900. Generally speaking, the more elaborate they are, the more valuable, the very best being those of oval section with Webb crystal glass bodies, cut in the so-called "rock-crystal" technique (pages 49 and 52).

An unusual set of four amber glass decanters — a little bribe, since each decanter is engraved "E & T. Taylor's Orange Wine" or whatever. The original case has the arms of William IV engraved on the cover, and the case is stamped in the corners with the Rose, Thistle, and Shamrock; I do not know what happened to the Welsh! Now this set combines everything desirable — unusual colour glass, fine quality mounts, original case, Royal association, and an amusing *raison d'être*.

£500 – £600

An early claret jug — date 1838. The earlier ones often have corks instead of the hinged covers which later became general. This jug has beautiful amethyst-coloured glass, which approximately doubles its value.

£150 − £180

An exceptionally heavy and elaborate early Victorian wine jug. If you can imagine it without the figures around the shoulder it would be more typical. Incidentally, all-metal jugs or glass largely overlaid with metal are known in the Trade as wine jugs, those basically glass with metal mounts and handles as claret jugs. Pairs of all are uncommon, and command a premium of 2½ to 3 times the basic price.

£275 – £350 (one)

Another very good type of wine jug, circa 1845, in which the plain glass body is entirely overlaid with a cage-work of grape-vines.

£225 – £275

An extremely fine pair of frosted glass wine jugs with matching coasters,
circa 1845, with Royal associations.

£950 – £1,200 complete
Coasters only of this type
£200 – £250 pair

A most unusual Edinburgh made wine jug with applied panels of dark blue enamel. Circa 1850.

£300 – £400

A so-called "Cellini" jug. Examples are found from about 1880, but are mostly after 1850. They vary greatly in quality, from cast ones superbly hand-chased, to die-stamped ones often made in Birmingham or Sheffield. As the quality, weight, size and date vary so greatly, I divide them approximately into three price ranges.

> *£100 – £135 (die-stamped, light, late)*
> *£135 – £200 (average, cast, or heavy die-stamped–*
> *£200 – £300 (early, cast and finely chased; or extra large)*
>
> *The goblets according to quality – per pair £80 – £125*

An unusual example by George Fox, circa 1865. He is a sought-after maker of excellent quality novelties.

£200 – £300

An ewer and basin of 16th century design. Although it *could* be hand-embossed, it is more probably reproduced by the electrotype process, by which an exceptionally heavy deposit of silver was plated into a female mould to form the article. Elkington were particularly noted for this type of work, and produced careful copies of many famous pieces. It has something of the feel and look of casting, but is very porous-looking at the back.

£350 – £450
£150 – £200 each piece separately

One of the most popular type of wine jugs. The body is usually frosted, sometimes gilt, and the handle twists round. It is a standard type of which quite a few must have bee made, around 1865.

£150 – £200

Although this has the same body shape as the previous example, it is
of a less popular pattern, with the bold sweep of the handle giving way
to typical Fern engraving. Same period to slightly later.

£110 – £150

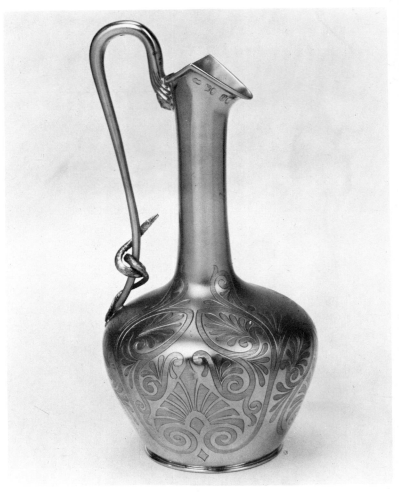

The Greek style — as early as 1850, but usually circa 1865. This is probably copied from a genuine example in pottery. Not madly popular.

£100 – £150

More Greek style — but rather finely flat-chased. Circa 1875.

£170 – £220 jug
£80 – £110 pair of goblets

One of those weird objects that appeals to me more than to most. It is a little Gothic, with touches of 16th century Spanish. Being by Stephen Smith, it is of excellent quality.

£75 – £100

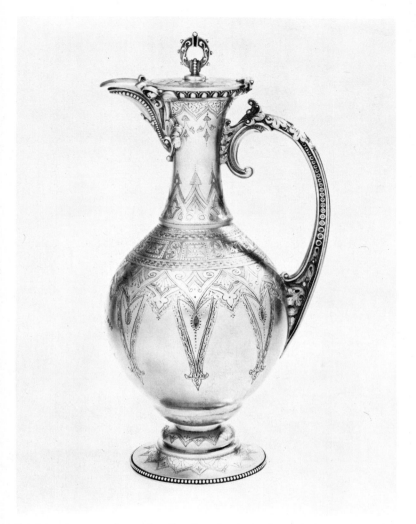

Bright-cut engraved, partly gilt, the silver part frosted. Circa 1880. Not a popular style, but the fine quality and condition is appealing.

£140 – £180

Claret jug in the Greek taste. Most claret jugs date from the 1880's and 1890's — this is a little earlier.

£55 – £70

A more popular type — with claret jugs, the more decorative ones seem to be preferred. Some of the best were made by W. & G. Sissons, of Sheffield.

£65 — £80

An attractive and decorative claret jug of fine quality. Note the "Japanese" influence in the glass of this and the following example.

£100 – £125

An outstanding example of rather unusual wide-necked form, the glass of Webb crystal, cut in a "rock-crystal" technique, probably by William Fritsche.

£200 – £250

Claret jugs in the form of birds or animals are occasionally to be found.
Beautifully made though it is, this example is too plain to fetch a really
high price.

£60 – £80

A good example of the 1890's. Heavy wheel-cut glass, elaborate mount and attractive form.

£90 – £125 (one)

A pair of claret jugs with oval section bodies cut in rock-crystal technique. This is a clearly defined type of jug, made 1890-1910, with silver bands round the bodies, elaborate mounts, and silver feet. They have always been expensive even when other claret jugs were cheaper than now. Pairs are 2½ to 3 times the price of singles with all claret jugs.

£125 – £175 (one)

A good quality, but ordinary cut glass jug. The lion and shield is a common finial. Circa 1890.

£60 – £80

Plain dull mount, reasonable cut glass. Very late.

£35 – £50

Perfectly plain glass on any of the preceding examples would probably reduce the values given by about 40%. Decorative examples are definitely preferred.

Elaborate William IV beer jug. Popular type.
£350 – £450

Large beer jug and cover, of a type popular about 1850. About 10 inches high. Not really such good quality as the next example; just compare the crispness and detail of the decoration.

£135 – £185

What the last one should be, but is not. Same period.
£200 – £275

This is an 18th century quart tankard and cover which has had a lip added, been embossed and had ivory insulators inserted into the handle, in the mid-19th century. Although basically Georgian, they would be of no interest to a collector of Georgian silver, and as they are found in considerable numbers they are best treated as 19th century objects. As such, they are quite sought after, most of them going to Italy, where they are sold as genuine examples.

£150 – £200

A large embossed water or beer jug. The sporting subject is particularly sought after, and is well done.

£225 – £300

An exceptionally good carved ivory tankard with Dutch or German silver mounts, circa 1870. About 12 inches high. Although not of British origin, quite a few of these tankards are to be found, although few with the fine bold carving of this one.

£500 – £650

If the ivory is badly cracked from natural causes, the value is considerably reduced, as it will be by accidental damage.

A fine copy of the late 17th century made in the late 19th century. Britannia standard silver was often used to imitate the original more exactly, and unscrupulous dealers have been known in the bad old days to remove most of the hall mark, leaving a badly worn Britannia and Leopard's Head, looking rather like the early marks.

£125 – £200

Another faithful copy — this time of a 16th century stoneware (or Tigerware) jug. Until recently, even 16th century examples were not very popular.

£65 — £80

Victorian wine-coolers are not too common, but are certainly amongst the most highly priced pieces of 19th century silver. As they are undoubtedly objects of great luxury, they were always well and solidly made; those who could not afford them made do with excellent plated examples. The pair shown above are unusually simple, especially for 1835. They are what the Trade call "ice pails"; that is to say that they have no liners (and never did have). The stands upon which they rest are also unusual. Wine-coolers today are used as often for pot-plants as for champagne. A single one is about one-third the price of a pair.

£800 – £1,000 pair

An extremely ornate example, with cast feed and handles. Note on the foot the grotesque face formed from a leaf; a decorative feature which goes back to the Romans. Although one may not care for them personally, lots of people will love them. This, by the way, is the normal type with a separate liner and cape (the bit at the top). Sometimes the liner, and the cape as well, if it is plain, are made in Sheffield Plate which reduces the value about 20%. Wine-coolers are not objects of everyday use, so they are normally in excellent condition. Otherwise one should be careful to look for holes in the elaborate embossing.

£1,000 – £1,250

An example in the Regency taste — again of fine quality with laid on
scenes.

£1,000 – £1,250 pair

A later example, about 1865, in the Greek taste. This is not particularly popular, although the standard of workmanship is high. They are also a bit vase-like.

£650 – £850

Wine-coolers in the form of the Warwick Vase were made throughout the 19th century. The quality of the modelling is important, and the value will be greater if the decoration is cast, as it probably would be on London ones, than if die-stamped as it usually is on Birmingham or Sheffield ones.

£850 – £1,100
Less 15% if less than 100 years old.

The use of silver mugs for beer drinking seems to have been on the decline in the 19th century. Tankards and covers are practically non-existent after 1810, except for highly decorative ones intended for presentation, and most Victorian mugs were intended for the use of children, and probably given as christening presents. These two are fairly typical of the William IV period, of vaguely thistle shape and standing on a separate stemmed foot. They hold about ¼ pint.

£35 – £45

The typical goblet of this period is also thistle-shaped and embossed,
6 to 7 inches in height.

£35 – £50

Two children's mugs of the 1840's.   £30 – £40

1 exceptionally fine example of the type of mug shown on the right
the last photograph. By comparison, it shows a finer and bolder foot,
th more depth and movement, and crisp and vigorous embossed
coration. It is also of pint size, a rarity.

£65 – £80

A rather better than average goblet. It looks like a Hunt and Roskell pattern, from the shape, and over-all grapevine decoration. Hunt & Roskell were the successors of Paul Storr and John Bridge, and held the Royal Warrant. Such goblets often formed Royal christening gifts, often with an accompanying knife, fork and spoon. A Royal association increases the value considerably.

£50 – £65

Typical goblet embossed with a hunting scene. About 7 inches.
£35 – £45

A half-pint mug. The shape is early George II, the decoration typically Victorian. An attractive design.

£35 − £45

As a change from most mid-19th century items, Robert Garrard (mainly) produced a few perfectly plain pieces. This child's mug was made circa 1850 and relies on its quality for its effect.

£35 — £45

Two further examples of this style. The goblet is hall-marked 1867, and the tumbler cup a few years earlier. Tumbler cups, which have a thick base and thinner sides so that they "tumble"(upright)when placed on their side, are not often found in the 19th century, and this is another example of an earlier inspiration.

goblet £30 – £40
tumbler cup £25 – £35

This is another copy, of the late 18th to early 19th century. It has never been a popular pattern, even if of the earlier period, but they are usually quite nicely made.

$£20 – £25$

Another child's mug, but of its own period, the 1870's. The beaded edges, plain loop handle, and single rather geometric engraving are points to note. It probably has a finely frosted finish.

£60 – £75

Matching goblets. Pairs are worth about three times one.
£60 – £75 pair

Rather nicely flat-based. Being more decorative than the previous pair, and probably heavier, they command a higher price.
£80 – £100

Goblets, fern-engraved, embossed, plain, and semi bright-cut. Six to
seven inches.

£20 – £35

A good cast fox head stirrup cup. Because, again, these are copies of Georgian ones they can be of any period. This, however, is top quality good weight, well-modelled, the coat well indicated and finished.

Over 100 years old £150 – £250
Not over 100 years old, but pre-1890, £80 – £100

Two spirit flasks, of 1887 and 1880 respectively. The former has a bayonet catch top, much more convenient than the ordinary screw cup with which the smaller is fitted. Both have cups which fit on to the lower half. In a quiet way, very saleable.

Larger – £25 – £35
Smaller – £15 – £25

The best type of heavy cast coaster is shown on page 35 with the wine jugs. These are of more ordinary type, although still attractive, made about 1850. Victorian coasters are quite uncommon.

£100 – £135

Calling the staff became more mechanised in the 19th century, so table bells are less in evidence. However, they are still popular when found. This is probably the best type.

£70 – £90

Unlike earlier bells, this one has a silver sleeve outside, with the bell proper inside.

£45 − £60

A Charles & George Fox folly — a pear-shaped bell.
£60 — £80

The mechanical shop-bell type — you twiddle the knob. The actual bell is inside with the works, and the outside is merely decoration, pierced and embossed.

£25 – £35

CANDLESTICKS

This section, which formed one of the biggest in a previous book, is one of the slimmest in this. The reason is, I suppose, the spread of gas-lighting amongst the wealthier classes, after which candlesticks went into such a decline that you hardly ever see a pair between 1830 and 1880. Highly elaborate candelabra continued to be made, however, and are much sought-after today. As always, the better class ones were cast, and the less good die-stamped. The parts for the latter were made in Sheffield or Birmingham, and were assembled there or in London. Cast ones were made in London. Some were hand-made in sheet silver, and are copies of earlier ones originally made that way. They are sometimes in Britannia standard silver, if the originals were.

Candlesticks tend to be used and cleaned a lot. If they are loaded, the relatively thin silver may wear through on the high spots, and some repairers fill these holes with soft solder, and plate over them. There is, however, no really satisfactory way of repairing them. It is also very difficult or expensive to remove knocks and dents in them, since they have to be unloaded, or even unsoldered into separate pieces to make good the damage. Such candlesticks, if damaged, should therefore be very cheap, and are not really worth buying. Prices are given for pairs — sets of four are worth somewhat more pro-rata, but not to the same extent as Georgian ones.

A typical pair of decorative loaded candlesticks, Sheffield made, 1830-1840. You can see how crisp and sharp these are, and this is the condition to look for.

£125 – £175

Although later, this candlestick is typical of the reproductions of 18th century cast candlesticks made, mainly by Robert Garrard, in the mid-19th century. The rococo style was also very popular. They are invariably of superb quality, and since they are very uncommon, are always highly priced. They weigh up to 80 ounces a pair.

£250 – £400 (about £5 per ounce)

A good pair of 4-light cast candelabra in the rococo taste, by Garrard. These are amongst the most popular of all items of Victoria silver.

£950 – £1,250 pair.

Some truly amazing examples were also made; some of these candelabra weigh 500-1000 ounces each, and are three or four feet high. Years ago they fetched 6d. an ounce over melt price, but today, when they come up, the price can be reckoned in thousands, or

about £3 per ounce

Incidentally, singles of four or more light candelabra fetch about one third the price of a pair.

A pair of cast London-made candlesticks of 1845. Even these are a reproduction of early 18th century German ones. But, as we have said, London-made candlesticks of this period are a rarity.

£220 – £280

This is a taper-stick, a little item about 5 inches high, used singly on a writing desk to seal letters with wax. Although, moderately common in the 18th century, they seem to have fallen into disuse in the 19th, except as part of an inkstand. This is a Harlequin taper-stick, a faithful copy of circa 1760, made about 100 years later. It is London-made and cast.

£90 – £120

Another copy of the 18th century, loaded, this time, by the well-known Sheffield firm of Martin Hall & Co. 1870/1880. About 10 inches ın height.

£45 – £60

Garrard again — copy of Charles II. made up in sheet silver of heavy gauge, not loaded.

£200 – £275

"Dwarf" candlesticks 4/5 inches high, always loaded, Sheffield or London hall marks, usually a variation on a Corinthian theme. 1880 - 1900.

£25 – £35

Another copy, of the Commonwealth period. Hand made from sheet-silver, not loaded. About 6 inches.

£60 – £75

Classical revival. 1880's. Cast, about 11 inches.
£125 – £150

large pair of 7 light candelabra; loaded candlesticks. The candlestick
a copy of a Sheffield pattern of circa 1770, or may even be from the
riginal dies, many of which were still in use at this period. Even
1ough of comparatively late date, 1880, these multi-light candelabra
·e so scarce and popular, that they fetch a very high price.
£800 − £1.000 pair

A pair of more normal domestic size; loaded candlesticks. Attractive style and simple proportion — about 18 inches high.

£250 — £350

small pair of oval loaded candlesticks, about 7 inches high, late 18th
entury style, 1890 – 1910. These, unusually, have strips of Mother-of-
earl inlaid up the columns, which makes them about £10 dearer than
rdinary ones.

£45 – £55

Another popular type, 1890-1910 – copy of circa 1755. Large, abov 12 inches, loaded.

£60 – £80 pair.

Chamber candlesticks are not very common either. This is an attractive pattern, about 1835, with decorative capitals and mounts. The nozzle and extinguisher should be hall-marked and conform with the chamber-stick.

£50 – £75 each

A "miniature" example, which is really a desk taper-stick, of vine leaf pattern, die-stamped in Birmingham.

£35 – £50

WAITERS AND TRAYS

Large numbers of waiters, or salvers, were made in the 19th century, and quite a number of large tea trays as well. They fall into remarkably few categories, however, the earlier ones being on the whole flat-chased with decorative borders, middle period ones engraved with somewhat less elaborate mounts, and the later ones usually bead edged with light engraving.

Waiter is the trade term for a salver; a tray is larger and has handles. Without handles, it remains a waiter, however large.

Waiters and trays are amongst the most abused items. They usually get intensive wear, but more serious is the removal of presentation inscriptions, monograms or armorials. Unless the article is of a very stout gauge, such removals usually leave the centre rather thin; by holding the article at an angle you will usually see a hollow at the front or the back and the centre will be wavy. Where the border comes down into the centre with points, check to see these are not worn through by cleaning. The junction between the border and the flat is sometimes broken and repaired. Anything with a worn or tatty appearance should always be avoided anyway.

A large early period tea tray 1830-1850. The gadroon and shell border shows Georgian influence, but the flat-chased decoration is more typical of the period. Trays like this are usually 150-170 ounces in weight.

about £3 per ounce.

Another good example with typical scroll and leaf border, and flat-chased decoration, somewhat less rigid than the previous example.
about £3 per ounce

Another good flat-chased tea tray. This particular shell-and-scroll border is known as a "pie-crust" border. It is a very common type. One would expect this tray to be a little smaller and lighter (125-150 ounces) than the previous example. Same period.

about £3 per ounce.

A large circular waiter with a cast border. The border is cast in sections and soldered on to the flat base. Owing to the thickness of the casting and the elaborate cast feet which they usually have, these waiters are rather heavy. They are usually, though not, invariably, of substantial size — 18 to 22 inches, weighing 100-150 ounces or more.

about £3 per ounce

An interesting Sheffield-made example which may look rather like the previous one; however, this is entirely the product of die-stamping. From the back, the border is hollow, and the edge is strengthened by being turned over. Even the flat-chasing on the front is stamped, although it may be hand-finished. These hollow borders are only found on Birmingham and Sheffield salvers, and are very likely to have holes in them, which really cannot be repaired satisfactorily.

£2.50 to £3 per ounce

A waiter of slightly later date — with a simpler border and engraved instead of flat-chased. Slightly less popular than the latter.

£2.50 to £3 per ounce

A smaller waiter with a plain shell and scroll border, copied from the mid-18th century. Engraved.

£2.50 to £3.50 per ounce

A later example still — 1860-1875. The feeling is less rococo, and a tiny bit lifeless, in spite of those swallows, which were such a popular feature.
£2 — £2.50 per ounce

A good tea tray of the same period.
£2 to £2.50 per ounce

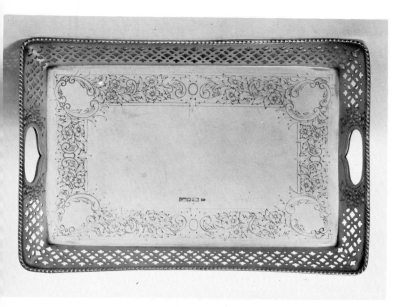

A rather unusual rectangular waiter with pierced gallery border. 1870's.
£2.50 to £3.50 per ounce

A very late waiter of the 1880's. Even less compelling with its perfectly round shape, and formalised fern and flower festoon engraving. The pierced edge is slightly unusual.

£1.50 to £2.50 per ounce

handsome turn-of-the-century large tea tray. Marvellous engraving.
bit too heavy for its own good, since anything other than a low price
er ounce will make it too expensive overall.

about £1.50 per ounce

A circular dish, which may have been accompanied by a ewer. Some examples are considerably more elaborate than this, with laid-on decoration and armorials. Fine examples, especially with some connections, can fetch very high prices; but for the average —

£2 – £3 per ounce

A wall plaque in the Aesthetic taste. About 12 inches in length. They do not seem very popular.

£60 – £75

VEGETABLE AND ENTREE DISHES

These are very popular items, since they remain as useful and decorative today as when they were originally made. Strictly speaking, an entrée dish is one in which the cover, by detaching the handle, can be used as a separate dish, whilst a vegetable dish has a domed cover with a fixed handle. Although you might think that entrée dishes are more useful, there does not seem to be much difference in price; possibly because without the design limitation imposed on entrée dishes by their dual function, vegetable dishes can be more attractive in appearance.

All serving dishes come in for a lot of hard use and abuse, especially, as sometimes happens, they have done duty in a Club or Regimental Mess. The bottoms tend to get badly cut up, and sometimes these cuts are polished out, leaving the bottoms thin. Arms are often removed from the covers, leaving thin spots in the sides. The top mounts of entrée dishes, where they sit upside down on the table, often are very worn from so doing.

Soup and sauce tureens and sauce-boats tend to have been erased, had the handles torn out by carelessness, or the feet pushed through by being banged down on the table.

With the exception of soup tureens and soufflé dishes most of the following items normally come in pairs. When there is only one, the price is about one third of that quoted.

A fine and decorative pair of vegetable dishes circa 1830-1850 with finials in the form of artichokes. Vegetable finials are always popular, but if detachable should be hall-marked to match the dishes as they are often improvements or replacements.

£400 – £550 pair

Typical oval entrée dishes with handles formed as crests of the arms. Circa 1830-50. You will notice they have matching warmers for the entrée dishes to sit on, and these are almost invariably in Sheffield Plate. Whilst they are of little value by themselves, they add £25 − £40 to the value of the dishes.

£325 − £425 pair

I cannot resist including these although it is unlikely that one will come across such a pair. They were made in 1846 by John Hunt, and weighed 700 ounces the pair.

£1,800 – £2,400

A vegetable dish and cover with its own stand and lamp, circa 1860.
£375 – £475 pair

A matching vegetable dish with a hot water jacket filled through the screw-off handle. They usually have double or triple dividers inside.

£400 – £525 pair

A very plain pattern of the turn of the century. I have always found these rather unsaleable, and so they are reasonable priced.

£1.25 – £1.50 per ounce

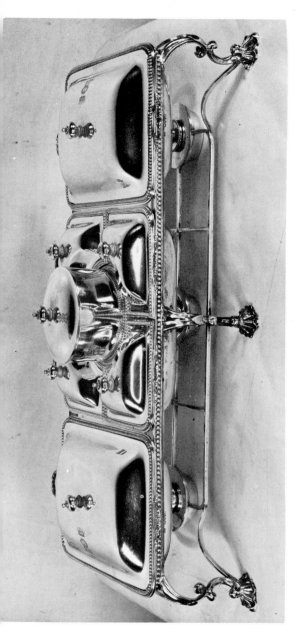

A most unusual supper set, circa 1880. It is difficult to put a price on such an item, as they are so rare and so sought-after.

£400 — £550

SOUP AND SAUCE TUREENS, SAUCE-BOATS AND SOUFFLE DISHES

Victorian soup tureens are quite common, and are also extremely popular. Very few of them are sold to residents of this country; they usually finish up in Italy or South America. Since these are the major markets, it is the more decorative examples which are the most highly prized. Sauce tureens and sauce-boats are far less common.

Soufflé dishes are made in two parts. They have a plain liner, usually made entirely in one piece, or at least hard-soldered, so that it can go into the oven without damage, and a decorative outer, into which the liner is placed for serving.

All these items may have the faults usually associated with hollow-ware, thin spots due to erasing, torn-out handles, and pushed-in feet. In addition, being very much articles for use, they may often be well worn. As far as hall marking is concerned, tureens should be fully marked on the body or foot, and partly marked on the cover.

Single sauce tureens are worth about one quarter the price of a pair. Pairs of soup tureens are worth only twice the price of one. Single sauce-boats are about one third the price of a pair.

This type of soup tureen dates from about 1830 to 1845 and is one of the most popular types.

soup tureen £850 – £1,050
pair sauce tureens £450 – £650

Middle period — 1845-1870. Also a good type, but less decorative than the previous example.

soup tureen £600 — £750
pair sauce tureens £300 — £375

Late Victorian boat-shaped copy of late 18th century original. As it is not to the taste of buyers and is less than 100 years old, it is not a saleable type.

soup tureen £225 – £300
pair sauce tureens £180 – £250

Highly elaborate sauce-boats, some George II influence, 1830-1850.
£300 – £400 pair

Another pair with rather elaborate feet and handles 1840-1860.
£250 – £325 pair

Sauce-boats on pedestal, or collet, feet, are less popular than those on three feet; however these have stands, which is unusual. But this border is not the most popular. About 1870.

£250 – £325 pair
without stands about half

Late Victorian reproduction — type sauceboat about 1800. Popular type, although late date.

£150 – £200 pair

About the best Victorian sauceboats you could find. All the decoration is cast and applied, a much more expensive process than embossing, which could have achieved an almost similar effect. These were made in 1880 — over 100 years old, plus 25%.

£350 – £450

A soufflé dish. It has a plain inner liner which can go in the oven, and is slipped into the outer case for serving. This is an unusually fine and elaborate example; the most common ones are quite plain, and often without feet.

this one £200 – £250
ordinary examples £2 to £3 per ounce

Dinner plates are in considerable demand abroad. They are usually in dozens, and any lesser number is of considerably less value, although eight is considered not too bad. Over a dozen, in sixes, they are *pro rata* with the dozen rate. Soup plates (not common in Victorian silver) are worth at least one third less. 9½ inches is the usual diameter; the larger they are over this, the better. This is the ordinary shaped gadroon, more or less unchanged since the early 18th century. All examples shown are assumed to be over 100 years old.

£750 – £900 dozen

A more elaborate pattern, and likely to be a larger size.
£800 – £950

Even larger, heavy, and more elaborate, therefore better, pattern.
£950 – £1,200

The most desirable meat dishes are the very large, over 20 inches, or the fairly small in pairs about 15 inches. The example shown is a large "treewell" or "venison" dish. These very large ones often have Sheffield Plate covers.

£2 – £2.50 per ounce

An unusual fish dish.   £2 – £2.50 per ounce

A seven bottle cruet. This type, with a number of glass bottles, became popular towards the end of the 18th century. Judging from the tens of thousands of plated ones existing, they must have formed the centre-piece of every middle-class Victorian family table. Silver ones, however, are uncommon. This is a particularly good example, with "hunting" scenes, made circa 1840.

£100 – £135

A small cruet; probably there would have been quite a number of these on the table — one per couple. About 1860, but too plain to be really saleable.

£25 – £30 each

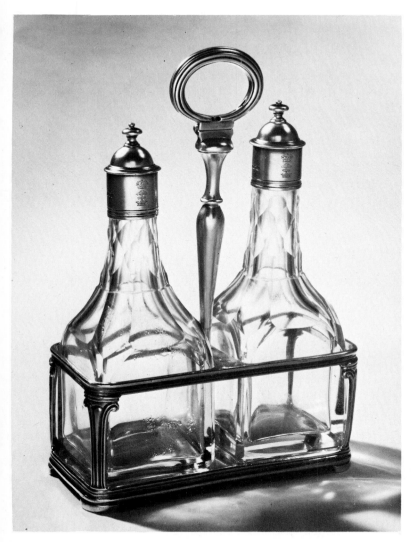

An oil and vinegar cruet copied from the mid-18th century by Garrard
about 100 years later. Excellent quality, and the square bottles are
unusual.

£85 – £110

Owl peppers and a mustard pot. The earliest examples I have seen are around 1840. The peppers are often rather thin and light; the mustard pots with the spoon terminating with a mouse in the owl's beak, are usually more solid. They are very sought-after.

Pepper £35–£45. Mustard £90–£150 according to size.

A cruet in the form of a newly-hatched chick (for mustard), the egg for pepper, and the broken shell for salt. The two chicks just hatching are a separate pair, but typical of many little bird and form peppers. All circa 1880.

cruet £80 – £100
pepper £50 – £70 pair

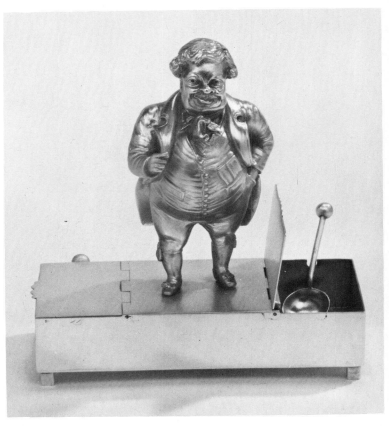

A Pickwick cruet. Two compartments in the base are for mustard and salt and the figure is the pepper, with holes pierced in the top of the head. There are pegs on the feet of the figure so that it can be lifted from the base for use. About

$£90 - £120$

The only complete Victorian condiment set I have seen, although there must be others. The individual pieces are typical of large numbers to be found so I have priced them individually. About 1870.

salts, pair £12 – £20
peppers, each £15 – £25
mustards, each £20 – £30

A typical larger, mid-period Victorian mustard pot. Somewhat worn decoration.

£20 − £30

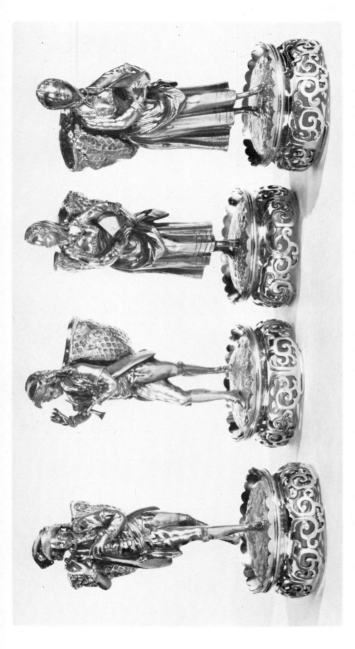

An unusual set of four figure salts, about six inches high, circa 1860. This sort of thing, being both highly decorative and useful, is highly valued.

£150 – £200 per pair

153

A large pair of salts with rams' heads. Quite large enough to be used for salted peanuts, which increases the value beyond that of mere salts. Again, highly decorative. About 1880.

£75 – £90

Early and mid-period Victorian sugar casters are unusual, mainly because of the use of sugar baskets, I suppose. This is a particularly good pair (and pairs, too, are rare).

£50 – £125 each
£125 – £300 pairs

A small bun-top sugar caster, about 1880, nicely decorated.

£20 – £30

Most late Victorian sugar casters are good reproductions of an earlier period, and this one is by Garrard, 1894. It is a solid piece of work, about 6 inches high.

£30 – £40

An "Adam" style example, slightly smaller, same period.

£30 – £40

I am not quite sure whether this is a butter dish or a muffin dish, but the former is more likely, so I include it in this section. About 1835.

£75 – £100

The more usual type of butter dish has a glass liner, and can be round or oval. They go from about 1835.

£50 – £75

An oval example, Sheffield-made, with a quite common type of finial in the form of a reclining cow.

£50 – £75

As I was captioning this, it occurred to me that it probably started
as a biscuit box with a taller cut glass liner, as the wheat-sheaf fini:
often associated with biscuit boxes! However, it makes a most acc
able butter dish, and I could be wrong anyway.

£50 – £75

An early period egg cruet for 4, with spoons.

£60 – £75

A later period example (you can tell from the engraving) for six —
about 1865.

£65 – £85

A very rare egg cruet which may be found minus its innards). About 1865.

£500 – £650

A very plain Victorian toast rack — they often have a Gothic flavour. In these days of ungracious living, not too popular.

£15 – £25

This type is still desirable, however. Decorative, sporting references.

£50 – £70

A picnic set which all fits into its leather case. The bee-hive thing screws into three for the condiments, and fits inside the bowl of the goblet, which fits inside the beaker. The other pieces fold, the one next to the double-headed spoon being a corkscrew. Most ingenious. Circa 1870.

£100 – £130

168

This is the type of sugar basket which made casters so scarce. This is not a very good type — too plain and light-weight. About 1850.

£35 – £45

Slightly better. Same period.

£45 – £55

Much better. The applied grape-vines can be either stamped, or cast, which is a more solid way of doing it.

£60 – £85

With a cover. All die-stamped in Sheffield.

£80 – £100

The best. All cast and London-made.

£100 – £135

Here we see the sugar basket turning into a sweet dish (for which they are all now used anyway). About 1860.   *£60 – £75*

Another Garrard copy of a 1790-ish dessert dish. Oval, about 6 inches long, the liner also in 18th century style.

£50 – £60

Three cake baskets, 1830-1860. The pierced ones are more popular and unlike Georgian baskets, Victorian ones are almost always round.

the smaller, lighter ones £50 – £75

larger, heavier, pierced £75 – £125

Two more, the one on the righr far more desirable, though both are excellent quality.

£75 – £100
£100 – £125

Late type, sometimes stamped, sometimes hand-embossed and pierced. Cast pierced foot. A popular variety.

£65 – £85

178

A flat fruit dish of very fine quality, about 10 inches across.

£80 – £100

A small sweet dish, about 5 inches across, in very crisp condition, using the same scene as that on many snuff boxes, no doubt struck from the same die. Although the surround looks hand-chased, it, too, is quite likely die-work. Birmingham made, of course.

£40 – £50

Small round baskets are often found with panels stamped with a number of these scenes, and with one at the bottom. I have known people count the number of scenes and assess the value of the basket at so-many-times what they think each one is worth, which usually brings the price to around £300! However, I have had several, and in my experience they are worth

£70 – £90

LARGER DISHES, BOWLS AND CENTREPIECES

A rather overpowering and heavily embossed fruit dish about 1840.
About 12 inches across and a very popular type.

£250   £350

A very attractive pair of dessert stands or comports with their original partly-frosted cut glass dishes. A little Chippendale influence, I would say. Circa 1850.

£180 – £250

A handsome Victorian oval comport on foot, about 15 inches across.
Circa 1870.

£200 – £275

A set of three comports, with their original dishes. Circa 1860. Although decorative, the silver weight is not great, so the price is less than one would think.

£250 – £350

Two comports, one with three figures, the other with a single one. Figure supports are popular, but the one on the left is definitely the more valuable and attractive.

£150 – £250
£125 – £175

An embossed comport on low foot.

£75 – £90

A very fine quality oval fruit bowl – in Dutch 18th century style. About 15 inches across.

£250 – £325

187

This is an exceptionally good embossed punch bowl, which could be anything from 1835 to 1870, or even later. The price assumes it is over 100 years old – otherwise 25% less.

£250 – £350

A not particularly attractive bowl. Half-fluted decoration of this type is never popular.

£2 to £2.50 per ounce

Late 19th century copy of a Queen Anne Monteith. Not a really faithful one, which would be worth more.

£3 – £4 per ounce

Somewhat similar, without the rim.

£2.50 to £3 per ounce

The slightest trace of Art Nouveau in the feet. About 18 inches across.

£2.50 to £3 per ounce

his is Dutch or German, and is die-cast. The style is instantly recog-
sable, but to be sold legally in this country, they should be English
ll-marked. Most of them will be found to be 1890-1910. As they are
ry decorative, and nothing of this sort was made here, they are quite
opular amongst those who like that sort of thing.

£3.50 to £4.50 per ounce

PRESENTATION CUPS

These vary from the most simple to the most elaborate. They were generally given as prizes; more imposing testimonial pieces usually took the form of large centrepieces or candelabra, often suitably modelled. These cups were given at point-to-points at racecourses, for yachting and so on. Obviously few of them have suffered much from use, the main danger to their condition being the removing of inscriptions so that they could be re-sold. Unless very carefully done, this could leave the cup very thin, especially, as some have been erased and re-engraved more than once! The more applicable a cup is to present day use, the more saleable. Horse-racing is still going strong — a cup embossed with a pony and trap (no doubt for the best equipage at some long-forgotten show) is obviously less so, since most cups are still bought for re-presentation.

ypical early cup, 1830-1840. It has the popular shape and late Regency ≥coration of the period, which is elaborate without being overdone. ote the border with the National Emblems of the United Kingdom. his cup has a hare-coursing scene, but being only about 9 inches high, more likely to be used for flowers than as a trophy.

£75 – £100

milar but much larger cups with covers of this period.

£2 – £2.50 per ounce

Highly decorative middle period cup and cover. Fine quality and workmanship make it desirable.

£1.50 – £2.50 per ounce

The type that is not desirable. As I have said, these Greek-type things are not popular unless very fine. Whilst of good quality, this example remains ordinary. Circa 1870.

£1.25 – £1.75 per ounce

The pony and trap type. Circa 1870.

£1.25 – £1.75 ounce

The one on the left is worth more than the one on the right, being more decorative.

£1.25 – £1.75

The one on the left is again worth more than the one on the right, being more pleasing in appearance, in the George II style, and having laid-on cast trap-work, a more desirable form of decoration than simple embossing. Also, large cups with covers are quite difficult to find, and are saleable.

£2 – £2.50 per ounce

The vase can be found from the reign of George III onwards. Like the Warwick vase, it is a copy of the antique. Most examples, however, are Victorian or Edwardian. They are usually about 18 inches high, and correspondingly heavy.

Over 100 years old	*£3.50 per ounce approx*
Under 100 years old	*£2.50 per ounce approx*
Post 1900	*£1.50 – £2 per ounce*

MODELS AND DECORATIVE OBJECTS

The Victorians were very fond of portrayals of animals, as can be seen from the tens of thousands of bronzes imported into this country from France, and from the stuffed fish, birds and animals that haunt a thousand antique shops. Silver models are relatively rare, since most groups were made by casting, which used a lot of silver, and was expensive.

Towards the end of the century, large numbers of Dutch and German models were imported. Birds, model boats or "Nefs" as they are called, wager cups and grotesque figures were among their specialities. They can be found with English import hall-marks from about 1880 onwards, the best known maker being Bernard Muller of Nuremburg. These have his initials, plus a mark like a pinecone, plus the letter N. usually.

A fairly early type of model, circa 1845. This is quite small, and the panniers may contain ink-wells or condiments.

£200 – £300

A large model after a bronze by that most prolific French sculptor, P.J. Mêne. About 1860.

£2.50 – £3 per ounce

A superb race-horse model. Look at the sheen on the coat!. This type of finish can only be achieved by weeks of hand-finishing. And, of course, this is about the most popular subject possible.

£5 – £7 per ounce

"Killing" subjects are not popular in this country, although they do not dislike them in America. This life-size model would not be easy to sell in spite of its fine quality. Circa 1880.

£2 – £2.50 per ounce

A charming decorative group — if you like that sort of thing. Beautifully
modelled. It is a pity that they usually bear only the silversmith's mark
and are not signed by the sculptor. Circa 1880.

£2.50 — £3.50 per ounce

Continental models which, unlike the English ones, are mostly chased up from the flat, with a consequent much lighter weight.

£4 – £6 per ounce (the lighter the dearer pro rata)

selection of Continental model birds.

£4 – £6 per ounce

A group of musicians with nodding heads. Very popular.
£6–£8 per ounce

he same sort of thing. The heads are on a spring, and come off at the
eck – ruff. Presumably you could keep something inside but I don't
now what.

£6 – £8 per ounce

A wager cup — usually female figures. This one is unusual. They range in height from about 4 inches to about 12 inches and are very sought-after.

4 inches £30 − £40, 8 inches £70 − £100, 12 inches £100 − £150

A small (about 12 inches) and comparatively simple nef.

£5 – £7 per ounce

A large and elaborate one — 18 to 24 inches.

£4 – £6 per ounce

A dressing table mirror of typical form, about 18 inches high. The silver stamped from a thin sheet, and pinned to a velvet covered wood frame. Being so flimsy, the silver is often broken. They are difficult to clean and repair, so one in poor condition is worth much less than the given price.

£60 – £100

A better type, which it is possible to clean without getting the under lying velvet covered in metal polish. Still very thin silver, die-stamped Assuming the same size as the previous example.

£80 – £120

A smaller, but much finer mirror, individually modelled and cast, and also of earlier date, circa 1870.

£100 – £150

A clock, die-stamped in thin panels. This one is quite large, about 12 inches high, and has a striking movement (as you can tell because it has two keyholes). Such elaborate silver clocks are quite unusual, and saleable.

£125 − £175

A silver clock about 4 inches high — timepiece movement. Circa 1900.

£35 — £50

A silver and enamel strut clock — Edwardian. These are surprisingly reasonable, and usually rather pretty.

£25 — £35

SNUFF-BOXES AND VINAIGRETTES

Box-making is a specialised job, and the centre of box-making in the 19th century was Birmingham. The best known maker is undoubtedly Nathaniel Mills, who worked from about 1820 to 1870, but that is not to say that his work is necessarily better in quality than others. Generally speaking, the larger and more elaborate a box is, the more valuable it is, apart from considerations of unusual design. The best snuff boxes have floral mounts raised up to ¼ inch from the surface, with convex sides boldly embossed. Engine-turning and reeding, the latter especially on the sides, is also most common. The greater part of the decoration is, however, die-stamped, the box-maker's skill consisting of the assembly, especially of the hinge or joint, in which the divisions of the knuckles should hardly be visible.

One of the most valuable types of box or vinaigrette is that which has a scene of some sort in relief on the cover. Some have sporting subjects but the so-called "castle-top" are most usual. These have views of well-known buildings such as Windsor Castle, Abbotsford (Sir Walter Scott's home), the Scott or Albert Memorials, etc.

Unfortunately, many of these boxes have been faked at a time when plain boxes were in little demand, and it was also possible to buy card cases with the same die-struck scenes for next to nothing. The scenes were cut out and soldered on to the covers of the boxes. There are two giveaways to this − firstly, if genuine, the edge of the plaque fits *under* the amount and is held in place by it, usually without solder. Fakers cannot usually be bothered to lift the amount and do the job properly, and the plaque joins the amount edge-to-edge, and this can be seen under careful examination. Secondly, boxes normally have a small plain shield for engraving on the cover. In the case of a genuine scene-top box, this plain shield is normally to be found underneath the base instead, since it could not be in its usual position. Obviously such a shield will not be found underneath if the scene has been added later. A box *may* be genuine without the shield underneath, but it is worth while being suspicious about it. Sometimes a plain lid is straightforwardly embossed − this was never done with an old box, which always presents a plain surface inside. Another dodge is to remove the original inscription from the cover, and replace it with an engraved scene. Engraved scenes in general are seldom genuine.

Boxes tend to be in poor condition from knocking about in the pocket — the amounts, corners and decoration should be crisp and sharp, the inside gilding hard, lustrous and reddish in colour. A hollow in the shield or top, a blurred line round the shield, or a hollow inside the cover, indicates an erased inscription. If well done, the value is not affected by erasure — in fact, until very recently inscriptions were invariably erased unless they were of exceptional interest. However, collectors and dealers are getting slightly more enlightened, and original inscriptions are usually left. A later inscription will almost certainly be removed, and detracts from the value if present.

A page of vinaigrettes —
Narrow floral mounts, better quality £30 – £45
"Book vinaigrettes", according to size £50 – £100

Small, light, engraved or engine-turned, unmounted £25 – £35
Heavy floral mounts, still better quality £45 – £65
"Castle-top" vinaigrettes £125 – £175

An excellent snuff-box with heavy mounts. Value also depends largely on size, especially as cigarettes are getting long and longer. This is a large box, about 4¼ inches across. Some dealers call them "table" boxes (as opposed to "pocket" presumably).

£250 – £350

Floral embossed convex sides, bold top mount. A very popular type of box.

4 inch size £250 – £350
3 inch size £175 – £250
2 inch size £100 – £150

Quite good type of floral mounted snuff-box.

4 inch size £150 − £250
3 inch size   £90 − £150
2 inch size   £40 − £70

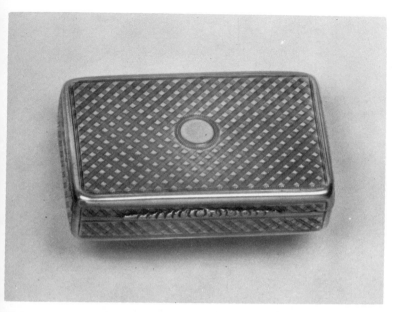

Plainer, unmounted type.

4 inch box £125 – £175
3 inch box   £75 – £100
2 inch box   £30 – £40

This has never been a popular shape, although the larger ones hold cigarettes. Note how you can tell the shield has been erased because the engine-turning around it is polished away, although otherwise the box is in excellent condition.

4 inch box £60 – £80   3 inch box £30 – £60
This shape is not usually found smaller.

A sporting snuff-box. These have hunting, shooting or hare-coursing scenes applied in the same way as castle-top boxes and vinaigrettes, and one should be equally wary of them. You cannot usually be as certain of their rightness as you can with this one, where the scene is hall-marked as well as the inside of the cover. Boxes on which these scenes are *engraved* are usually ordinary ones from which inscriptions and/or engine-turning have been removed, and the engraving substituted. Very fine sporting boxes sometimes have further scenes round the sides and bordering the base. Castle-top snuff-boxes would be roughly similar in value.

1½ inch £90 – £125   2 inch £125 – £175   3 inch £225 – £300

228

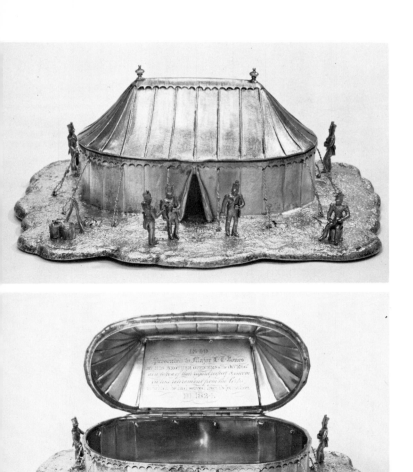

Rarity. Large and exotic table snuff-boxes of this type are always expensive.

£400 — £500

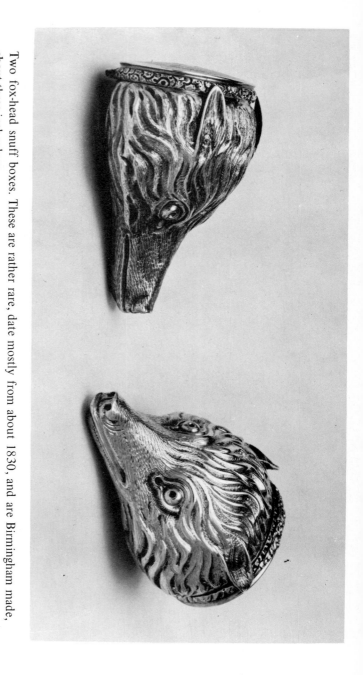

Two fox-head snuff boxes. These are rather rare, date mostly from about 1830, and are Birmingham made, about three inches long.

Price Range: *£250 - £400   if pre-1830 plus 30%*

Horse's hoof snuff box — usually about 1880.

£35 – £50

CARD CASES

These are popular collectors' items, since they offer a wide variety of type and price, and provide on a small scale examples of most types of decoration. Large numbers of embossed scene card cases have been cut up to provide the plaques for castle-top snuff-boxes, which is why they are in fairly short supply. The majority of embossed scene cases have the scene on one side only; the other is embossed all over with scrolls, with a small central cartouche for engraving. If both sides have scenes, the value is increased about 25%. Scenes, especially in high relief, easily get worn. To be desirable, cases should display crisp detail, as in the examples shown here.

An early embossed case of plain outline, with a rather unusual view of Windsor Castle. Circa 1840.

£35 – £45

The most typical type, with a scroll edge. View of the Scott memorial in Edinburgh c. 1845-1865.

£35 – £45

Exactly the same — but engraved instead of embossed. Much less desirable. Same dates.

$£20 - £28$

An Elkington electro-type, partially gilt. Rare. Circa 1860.

£50 – £65

All-over engraved, no scene, rather coarse. This type was made until the end of the century.

£14 – £18

Japonaiserie — this is not a very fine example, and is priced accordingly.
Good parcel-gilt examples can fetch three times as much.

£15 – £20

An end-of-century variation is the type opening like a cigarette case. They are usually very light-weight, and silk-lined with several compartments.

£10 – £12

LARGER BOXES

These are uncommon in Victorian silver, except towards the end of the century. The casket is typical of the finer quality productions of the middle of the century. Tea caddies are unusual, and often copies of the 18th century — I have shown one which is really Victorian. Thousands upon thousands of plated biscuit boxes were made; silver examples are uncommon. In the 1880's and 1890's large numbers of German and Dutch made cigar-size boxes were imported. They are usually of light gauge, die-stamped or, rarely, hard-embossed with tavern scenes, etc. English freedom boxes are also found, and can be recognised from their characteristic shape — long and narrow to take a scroll. They can make quite attractive cigar or cigarette boxes if suitably lined, but are very unsaleable in their original condition.

A very attractive casket by Robert Garrard — about 8 inches across. Circa 1860.

£250 – £350

Victorian tea caddy circa 1850. Typical lumpy style.

£100 – £150

Oval silver and cut-glass biscuit box 1874. Attractive and saleable.

£125 – £175

This is a folding biscuit-box in a half-open position. It is a triple-shell; most open in two. The pierced grilles inside hold the biscuits separate when it's closed. Very unusual in silver.

£200 – £275

Late 19th century Dutch cigar box with "Teniers" subjects (peasants and all that). It is unusual to have a lock.

£50 – £100

An unusually elaborate example, with cast figure feet, and cast applied mounts

£100 – £135

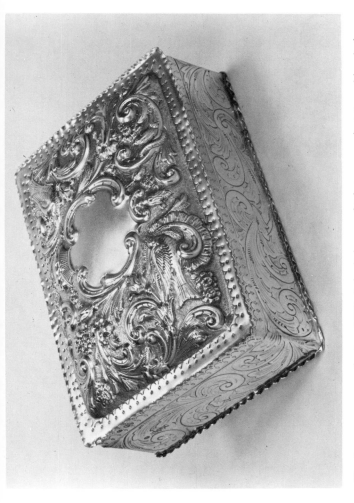

The home-grown product – 1888. I don't know which is worse, but it could only have been so crudely made in an attempt to compete in price with the Continental boxes.

£50 – £75

A Continental toilet box, about 4 inches.

£25 – £30

A good quality English example with cast top, well modelled and chased. 3 inch diameter.

£30 – £35

A jewel box in the form of a small table, the top embossed with the ever-so-popular at this time angels' heads. This motif can be seen on every type of article for the dressing table from 1895-1915. The drawer is velvet-lined.

£25 – £35

INKSTANDS

These were made in large numbers, and must have graced most desks. They range from delicate little items for a lady's boudoir to massive ones for the boardroom table. Of course, in those days before the fountain and ball pen, they were absolutely necessary.

Although few people use inkstands for their original purpose, they still form a highly decorative accessory for the desk, and are extremely popular. They have lots of separate parts, each of which should be hall-marked; base, gallery (if it has one), bottle-holders, mounts and tops, stamp box and cover, or (if fitted) taperstick with nozzle and extinguisher. If the bottles don't fit, they are probably not original, especially if the mounts aren't marked, and definitely not if they have a different mark from the base! Quality varies a lot, from stamped out unmounted examples (which owing to the die-cutter's skill can look quite impressive until you pick them up) to the very fine example on page 255 with heavy cast feet and mounts. Price is also affected by size.

An unusual silver-mounted William IV or early Victorian shell inkstand. It is about six inches high. The wire scrolls hold the pen.

£30 – £45

A shell-shaped inkstand with an "Aladdin's lamp" inkwell. Single-well inkstands are not too popular; apart from this example, they are usually circular, with a central glass inkwell with a silver top. Circa 1860.

£50 – £70

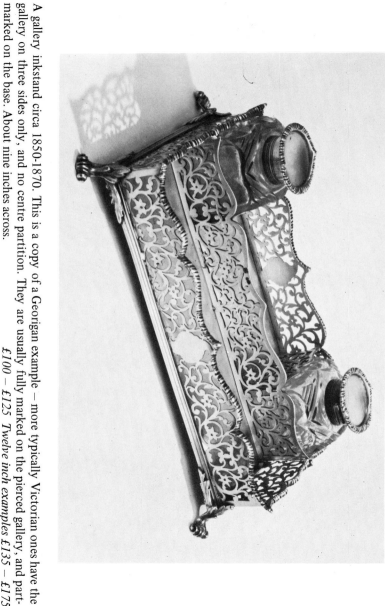

A gallery inkstand circa 1850-1870. This is a copy of a Georigan example — more typically Victorian ones have the gallery on three sides only, and no centre partition. They are usually fully marked on the pierced gallery, and part-marked on the base. About nine inches across.

£100 – £125 *Twelve inch examples £135 – £175*

A very fine example with finely cast and chased feet and mounts. Circa 1850. This type often has a chamberstick on top of the centre box. Circa 1850 – about twelve inches across.

£225 – £275 (without the later inscription!)

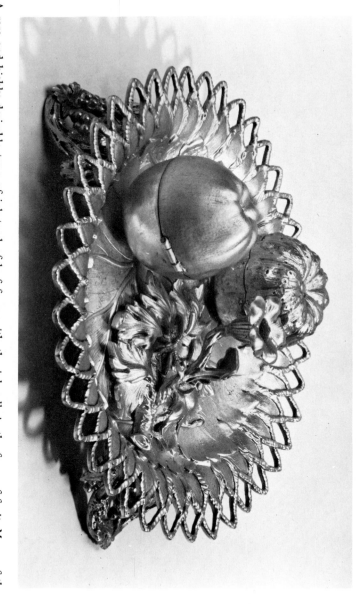

A rare and highly desirable type of inkstand of leaf form with the inkwells in the form of fruit. Many of these were made by Charles George Fox, and they are usually of fair size — about 9 inches across.

£375 – £525

A small mid-century example, before the scroll decoration gives way to the more geometric lines of the later examples. About 6 inches across, and of quite light construction.

£60 – £80

A later type — lighter in weight, too. About 1880.

12 inches £100 – £135

8 inches £80 – £100

A small pierced inkstand of about 1880. Being by the last of the Fox's it has considerably more character and quality than the rather ordinary example which follows.

£50 – £80

A small pierced inkstand — late Victorian. With all pierced inkstands it is important to check that the piercing has not been broken.

£25 – £35

A "treasury" inkstand — the glass inkwells hidden below the hinged covers. Sometimes they have hinged handles at the sides, and some can be very large and heavy. Turn-of-the century. This example is about eight inches across.

£80 – £100

MISCELLANEOUS

An incredible variety of small silver items were made in the 19t
century — it would be impossible to show even an appreciable pe
centage of them. However, I show a few for which there is some co
lecting interest.

This brandy saucepan is a copy of a Georgian one. They are quit
uncommon in Victorian silver, and may fetch from *£25 to £10*
according to size and weight.

Child's rattle — early 18th century examples exist, but most date from the Victorian period. This is not a very good example — it is small and light, with a thin piece of coral. Incidentally, don't be tempted to buy one with the coral missing — it's almost impossible to replace it these days.

£18 – £25

Two better examples — much heavier and bolder, one with two rows of bells. As these were not normally soldered, they are often missing, and replacements are hard to find. Common makers of rattles were George Unite and Yapp & Woodward. They are almost all Birmingham made.

£35 – £55

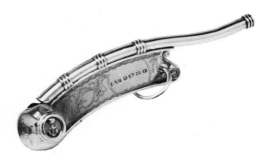

Boatswain's (pronounced bo'sun's) whistle — used to pipe officers aboard ship. Mostly 1850-1880, although Georgian examples are found. Same makers as rattles.

£30 — £45

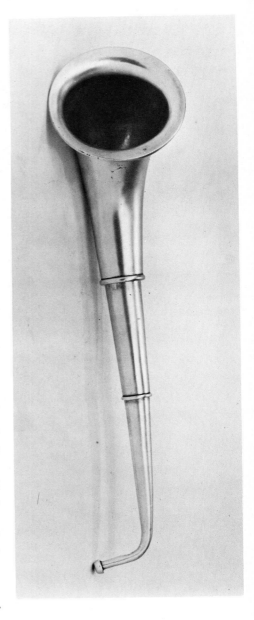

Ear-trumpet. This one is in three telescopic sections, but many varieties are found, often with the maker's name and patent. As with most "medical" antiques, there is quite a strong demand for these.

£50 – £125

Scalpel case — containing two or four very sharp little scalpels usually with ivory, tortoiseshell or Mother of Pearl covers. Middle of the century.

£30 − £45

Two silver-mounted scent bottles. These were mostly made about 1880 by a firm called Samuel Mordan & Co., of London. Their mark SM is not in Jacksons. All their things are beautifully made. The "lamp" is a scent bottle — the panels are of ruby glass and the conical top unscrews. The other has a screw cap at one end, a snap-closing cap at the other with spring sealing, and a small compartment in the centre. The caps are set with turquoise.

£25 – £50

ART NOUVEAU SILVER

I cannot go at length into the development of this movement. which has been very well dealt with by a number of authors since the style became fashionable a few years ago. Basically, the movement developed from a desire to get away from the excesses and ugliness of popular taste of the 1850's and 60's which, together with the development of machine-made articles, led people of taste to revolt against them and return to simpler, "hand-made" things. The pre-Raphaelites, the Aesthetic movement (which was strongly influenced by Japanese taste), the William Morris school, were all developments of this feeling. These were at all times minority tastes and, because of their deliberate attempts to oust the machine, mostly rather expensive, so the best and most characteristic examples were quite uncommon, and their owners confined to that small class of society with both money and taste.

However, later on the Art Nouveau style became rather commercialised, and many very inexpensive items were produced, die-stamped in very thin silver, which can nevertheless be quite attractive. There were always a number of "artist-craftsmen" who designed and hand-made their own productions – some of them, like Omar Ramsden, continuing right up until the second world war.

The objects which are most sought after today are those by well-known designers such as Christopher Dresser, C.R. Ashbee (usually hall-marked 'GoH' for the London Guild of Handicrafts), Nelson, Dawson, Ramsden, or Ramsden & Carr. The firm of Liberty & Co. (mark 'L & Co.' in three overlapping diamonds, and with Birmingham hall-marks) was foremost in popularising Art Nouveau – in fact abroad it is sometimes known as "Style Liberty". Their productions vary greatly in quality. Not too much is known about their designers, whom they did not publicise.

Most dealers do not like or understand Art Nouveau silver, although most of them now have some idea what it looks like. It is usually possible to buy pieces quite reasonably from the average silver dealer. Once pieces get into the hands of specialist Art Nouveau dealers, prices seem to rise out of all reason, and it often doesn't seem to make a lot of difference to the price if the article is silver or plated. I would advise collectors to visit these people for their education, but avoid them for their purchases!

A small, quite commercial production by the firm of William Hutton & Son, of London & Birmingham. Although the body motifs are die-stamped, the handle has that "hand-wrought" look, rather reminiscent of wrought-iron. The use of rivets is another typical practice.

£20 – £25

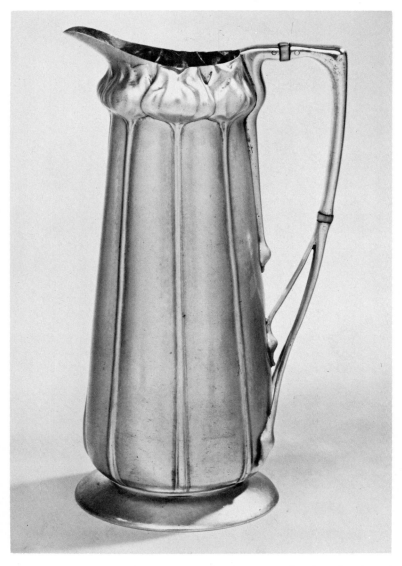

One of Liberty & Co's better quality productions — hand-made of good gauge silver. About 9 inches high.

£100 – £150

A small capstan inkstand by a follower of Ramsden — F.S. Greenwood.
In fact, it is signed like Ramsden's pieces "F.S. Greenwood me fecit".
It is in Ramsden's "Mediaeval" style with briar roses. The hammered
finish could only be done by hand, and was therefore popular.

£25 – £45

A condiment set by Ramsden & Carr, of 1912. This is in the Art Nouveau, rather than medieval style, all swirly and curly. This is a good size set, of excellent quality.

£100 – £150

An outstanding example of Ramsden's work — Art Nouveau styling, especially in the feet and corners, with fairy-tale Gothic landscapes and verses from "The Lady of Shallott". The cover plaque is enamel. Larger pieces by Ramsden are comparatively rare, and the enamel is very desirable;

£300 – £400

wo attractive pieces of no great value so far. The dish is by W.G.
onnell (the mark is in a clover-leaf stamp) and the case by the Gold-
niths & Silversmiths Co. Although well made, they are "commercial"
roductions. Both about 10 inches.

£50 – £75 each

A claret jug in the style of Dr. Christopher Dresser. He was a design
well in advance of his time, and pioneered an extremely plain a
geometrical style as far removed from the general style of the period
could be. Most of his work was around 1880, and most of it was ma
by Hukin & Heath (mark H & H) in silver and plate. Some was a
produced by Elkington. To be really desirable, pieces should be
Dresser's stamped signature. The jug is in his *style;* with his name
would be five times the price or more

£25 – £35